我爱中国菜 素食篇

Easy Recipes Easy Chinese

Vegetarian Dishes

First Edition 2015
Third Printing 2024

ISBN 978-7-5138-0929-0
Copyright 2015 by Sinolingua Co.,Ltd
Published by Sinolingua Co., Ltd
24 Baiwanzhuang Street, Beijing 100037, China
Tel: (86)10-68320585, 68997826
Fax: (86)10-68997826, 68326333
http://www.sinolingua.com.cn
E-mail: hyjx@sinolingua.com.cn
Printed by Tangshan Xicheng Printing Co., Ltd

Printed in the People's Republic of China

前言

中国菜，是世界上最错综复杂的烹饪体系，令人眼花缭乱，神往不已。

中国菜，以档次食材，可分为家常菜、市肆菜、寺观菜、官府菜、宫廷菜、药膳菜、民族菜等；以烹饪技法，可分为冷菜、热菜、大菜、小菜、甜菜、汤菜等；以风味流派，可分为鲁菜、川菜、粤菜、苏菜四大菜系；加上闽菜、浙菜、湘菜、徽菜，扩为八大菜系；加上京菜、鄂菜，是为十大菜系……林林种种。

中国人爱吃，懂吃，会吃，翻着花样吃。食材上可至鲍燕翅参，下可至根叶蔬果，可阳春白雪，可下里巴人。炒、烧、烤、蒸、煮、炸、煎、凉拌、淋，极尽烹饪之能事。中国人，是世界上最在乎口味的民族，极尽口舌之欲，孜孜不倦地追求着味觉的巅峰享受。对于任何食材，中国人就像一个充满了想象力的水墨画家，挥毫泼墨，巧妙地在素帛之上织锦添花，幻化出最奇妙的味觉之旅。

当下，素食以其健康的特点风靡世界。素菜也是中国菜的重要组成部分，本书呈现了一些简单易学的中国家常素菜，在保留蔬果、豆制品健康风味的同时，更配合中式烹饪特有的手法，激发出素食更丰富的味觉效果。

中国菜，讲究色香味俱全，所以中国人在做菜时会尽量避免单一色彩，绿色的葱花、辣椒、香菜或者其他香草，红色的干辣椒或者甜椒，黄色的生姜，白色的芝麻、大蒜，都可用来配合烹饪。第一可以增加菜肴的香气和风味，第二可以丰富菜肴的色彩，以此来增加菜肴的整体风味。

中式烹饪与西式烹饪在调味上有所区别，西式烹饪讲究定量，往往会精确到克、毫升或者勺，而中式烹饪更加尊重烹饪者的烹饪习惯，随意性强，创造性也更强，每个烹饪者的味觉习惯不一样，烹饪的手法和调味习惯也不一样。就像莎士比亚说的"一千个观众就有一千个哈姆雷特"，一千个烹饪者和品尝者就有一千种味道，所以本书中的调味品用量仅供参考。

中式烹饪讲究菜型的和谐，所以注重刀法，同样的食材或者香料辅料根据整体菜肴的造型会切成不同的形状，可切片、切块、切碎、切段。

生姜、大蒜、小葱，是中式烹饪不可或缺的香辛料，其中生姜最好不要去皮，因为生姜本身性热，而姜皮却是凉性的，可以中和生姜的热性；大蒜去皮后切碎或者拍裂后，应该先在空气中曝露一段时间再烹饪，这样可以使大蒜中的蒜素充分氧化，蒜香更浓烈，保健效果更好。

烹饪，本就是一项创造性的活动，在水火之间，在刀铲之下，食物涅槃，融入烹饪者的心性，化为倾倒众生的尤物。愿大家在庖厨之间，感受中式烹饪的魅力，也感悟蕴含其间数千年博大精深的中华文化，谢谢。

Preface

Chinese cuisine is one of the most complicated ones in the world. It is often presented in a dazzling way, so people are always yearning for it.

In terms of the ingredient quality, Chinese dishes can be categorized as home-style dishes, gourmet dishes, vegetarian dishes, official residence style dishes,royal dishes, medicinal foods and national dishes. From the perspective of cooking skills, Chinese cuisine also includes cold dishes, hot dishes, main courses, side dishes, sweet dishes and soups. Chinese dishes can also be further divided into ten types based on their flavours: Shandong cuisine, Sichuan cuisine, Cantonese cuisine, Suzhou cuisine, Fujian cuisine, Zhejiang cuisine, Hunan cuisine, Anhui cuisine, Beijing cuisine and Hubei cuisine.

Chinese people love to eat and are particular about their cooking methods so they often come up with different ways to cook. The ingredients range from abalones and sea cucumbers to vegetables and fruits while the cooking methods include stir-frying, braising, roasting, steaming, boiling, cold mixing and filter sprinkling. By selecting the ingredients and cooking method, one can cook both refined and everyday dishes. Chinese people care mostly about the tastes in the world and will tirelessly pursue them. A Chinese cook can be compared to a painter with a full imagination. He is adept at choosing the most suitable ingredients to cook a dish, just like the painter who excels in using the most appropriate strokes to create a wonderful painting.

Nowadays, vegetables, which are an important component of Chinese cuisine, are popular with people around the world who want to keep healthy. This book presents some easy-to-learn Chinese home-style vegetable dishes, which are conducive to becoming healthy, and showcases the rich flavours of Chinese dishes.

When cooking a dish, Chinese people pay attention to its colour, aroma and flavour, so they often include green scallions, peppers, coriander or fragrant herbs, dried or sweet red peppers, ginger, white sesame seeds or garlic. Because of this approach to cuisine, each Chinese dish often has its own unique feature.

In terms of seasoning, there are considerable differences between China and the West. Western cuisine focuses on specific measurements that are accurate to the gram, millimetre or ounce; however, Chinese cuisine can be much more individualized, impulsive and creative, and cooks frequently adjust their cooking methods, seasoning and ingredients according to their preferences. Just as Shakespeare said, "There are a thousand Hamlets in a thousand people's eyes", so we can say there are a thousand types of flavour if there are a thousand cooks or customers. Therefore, the amount of seasoning specified in this book is just for reference.

For Chinese cuisine, cutting and slicing skills are very important since presentation is one of the focuses. The same ingredients are often cut into different shapes, such as sliced, diced, minced or cut into sections to match with the style of a specific dish.

Ginger, garlic and scallions are indispensable ingredients in any Chinese dish. According to traditional Chinese medicine, ginger is hot in nature but its skin contains cold elements, so if we keep the skin, it can offset the hot elements; therefore, it is better not to peel the ginger. As for garlic, it should be peeled and cut into tiny pieces or cracked. Then it should be exposed to air for oxidization before cooking. This way, the garlic will emit a much stronger aroma and is also good for your health.

Cooking is an innovative activity, which can showcase each person's personality and creativity. May you become fascinated with Chinese cuisine, enjoy their charms and appreciate the profound and extensive Chinese culture behind them. Thank you!

Contents

Part 1 凉拌 / **Cold mixing** 1

刀拍黄瓜 /Smashed Cucumber 2

凉拌藕 /Cold Dressed Lotus Root 4

花生仁菠菜 /Spinach with Peanut Kernels 6

老醋木耳拌洋葱 /Wood Ear Mushroom and Onion Dressed with Vinegar 8

香辣海带丝 /Spicy Seaweed Strips 10

糖醋凤尾白菜 /Sweet and Sour Cabbage 12

时蔬大凉拌 /Cold Dressed Vegetables 14

小葱拌豆腐 /Tofu Mixed with Chopped Scallions 16

凉拌三丝 /Three Kinds of Mixed Vegetables 18

糖拌西红柿 /Mixed Tomatoes with Sugar 20

菠菜拌粉丝 /Spinach Mixed with Chinese Vermicelli 22

炝拌生菜 /Lettuce Dressed with Hot Oil 24

PART 2 热炒 / STIR-FRYING 27

家常豆腐 /Home-style Tofu 28

辣炒空心菜 /Stir-fried Water Spinach with Pepper 30

酱爆蔬菜丁 /Sauteed Vegetable Dices in Sauce 32

醋熘白菜 /Stir-fried Chinese Cabbage with Vinegar 34

清炒西兰花 /Stir-fried Broccoli 36

水芹菜炒香干 /Stir-fried Celery with Smoked and Dried Tofu 38

西葫芦炒鸡蛋 /Stir-fried Summer Squash with Eggs 40

酱烧茄子 /Sauteed Eggplant with Soybean Paste 42

VII

红糖蜜莲藕 /Preserved Lotus Root with Brown Sugar 44

香辣藕丁 /Spicy Lotus Root Cubes 46

糖醋青椒 /Sweet and Sour Green Peppers?........................48

荷塘小菜 /Stir-fried Lotus Root and Snow Peas 50

Part 3 小蒸 / Steaming 53

蒜泥茄子 /Steamed Eggplant with Garlic 54

剁椒蒸金针菇 /Steamed Golden Needle Mushrooms with Chopped Peppers 56

奶香玉米小南瓜 /Steamed Pumpkin and Creamy Corn Kernels 58

贡米酿藕 /Stuffed Lotus Root with Purple Rice 60

蒜蓉蒸丝瓜 /Steamed Towel Gourd with Mashed Garlic.........62

冰糖蒸山药 /Steamed Yam with Rock Sugar 64

剁椒蒸娃娃菜 /Steamed Baby Chinese Cabbage with Diced Hot Red Peppers 66

红枣酿苦瓜 /Stuffed Bitter Melon with Jujubes 68

PART 4 微煮 / BOILING 71

水煮玉米 /Boiled Corn 72

盐水煮毛豆 /Boiled Green Beans in Salty Water 74

五香啤酒花生 /Five-flavour Peanuts in Beer 76

砂锅白菜豆腐 /Simmered Chinese Cabbage and Tofu in an Earthenware Pot 78

金针菇豆腐汤 /Golden Needle Mushroom and Tofu Soup 80

菇鲜豆腐 /Mushroom and Tofu 82

蔬菜保健汤 /Vegetable 84

PART 1
凉拌
Cold mixing

Dāopāihuángguā
刀拍黄瓜
Smashed Cucumber

INGREDIENTS:

2 cucumbers
ginger, garlic and dried pepper to taste
10ml sesame oil
3ml mature vinegar
10ml sesame oil
3g granulated chicken bouillon (or MSG)
5g salt

TIPS:

Smash the cucumbers with the knife blade before cutting them into sections, so as to let in more flavour and strengthen the chewy texture.

Put some salt first and wait to rid the cucumber of its excess liquid. This will make the cucumber crisper.

Select cucumbers with flowers on the top and small thorns on the body; this indicates freshness.

Add flavour to the cucumber with the seasoning sauce before spreading hot oil, so as to make the dish tastier.

DIRECTIONS:

1. Wash the cucumber and remove both ends. Smash it with the knife blade.
2. Cut the cucumber into small sections.
3. Spread a small amount of salt on the cucumber sections and blend them. Put it aside for 15 minutes. Strain the excess liquid.
4. Mash the ginger and garlic, cut the dried pepper into small sections and remove the seeds.
5. Blend the mature vinegar, light soya sauce, sesame oil, bouillon and salt into seasoning sauce.
6. Blend the mashed ginger, garlic, seasoning sauce and cucumber sections.
7. Pour cooking oil into the pot and heat it up, stir-fry the dried pepper in the oil and spread it on the cucumber. Blend it and wait for 20 minutes.

EASY CHINESE:

qiàng
炝 fry quickly in hot oil

chǎo
炒 to stir-fry

dùn
炖 to simmer

shāo
烧 to braise

Liángbàn'ǒu
凉拌藕
Cold Dressed Lotus Root

INGREDIENTS:

2 lotus roots
1 red sweet pepper
ginger and green onion to taste
10ml light soya sauce
3ml white vinegar
3g sugar
10ml sesame oil
3g granulated bouillon chicken (or MSG)
5g salt

TIPS:

As lotus root contains starch, it is better to rinse off the sliced lotus root so that the dish tastes fresher.

Oxidation may take place when the lotus root is exposed to air, which will cause discolouration, so the sliced lotus root should be soaked in water with white vinegar to stop the reaction.

DIRECTIONS:

1. Wash the lotus roots and cut off the two ends. Peel it.
2. Cut the lotus root into thin slices and put them into a bowl. Rinse off the starch on the surface.
3. Soak the lotus root slices in water with white vinegar.
4. Mince the ginger, red sweet pepper and green onion.
5. Pour cooking oil into the pot and heat it up, stir-fry the chopped pepper and ginger and then blend them with the vinegar, light soya sauce, sesame oil, bouillon, sugar and salt to create the seasoning sauce.
6. Let the water trickle off the lotus root slices and put them back into the bowl. Sprinkle the seasoning sauce on the lotus roots. Spread the chopped green onion on top and serve.

EASY CHINESE:

Wèidào zěnmeyàng?
味道怎么样？
How does it taste?

Hǎochī jí le!
好吃极了！
It's delicious!

Huāshēngrénbōcài
花生仁菠菜
Spinach with Peanut Kernels

INGREDIENTS:

300g spinach
50g peanut kernels
Small red pepper and ginger to taste
10ml light soya sauce
10ml sesame oil
3g granulated chicken bouillon (or MSG)
5g salt

TIPS:

Make sure to boil the spinach before cooking in order to separate the calcium oxalate out and also for better tasting spinach. This is also conducive to the prevention of calcium stones and good for the assimilation of calcium.

Some vegetables are rich in chlorophyll. Adding a small amount of salt and cooking oil when cooking these vegetables in boiling water can make them look greener and more presentable.

When frying the peanuts, make sure to put them into the pot before the oil becomes hot. Scoop the peanuts up as the colour turns to a light reddish-brown, then allow the oil to trickle off. The peanuts will continue to slowly cook and become crispy due to the residual heat.

DIRECTIONS:

1. Pour a large amount of cooking oil into the pot and add the peanut kernels while the oil is still cool. Fry the peanuts slowly on medium heat until they are cooked. Remove the peanuts and allow the oil to trickle off. Allow the peanuts to cool.

2. Remove the yellow leaves and root of the spinach before rinsing it. Soak the spinach in lightly salted water for 10 minutes, remove and allow the water to trickle off.

3. Put water into the pot. Add a small amount of salt and several drops of cooking oil, then bring to a boil. Put the spinach into the boiling water. Remove after 2 minutes.

4. Soak the spinach in ice water.

5. Squeeze the juice out of the ginger and blend the ginger juice with the light soya sauce, sesame oil, bouillon and salt to make seasoning sauce.

6. Slice the pepper along the width, allow the water to trickle off of the spinach and mix it with the sliced pepper.

7. Sprinkle the seasoning sauce on the spinach and pepper, then mix them well. Spread peanuts onto it and serve.

Lǎocùmù'ěrbàn
老醋木耳拌
yángcōng
洋葱
Wood Ear Mushroom and Onion Dressed with Vinegar

INGREDIENTS:

10g dried wood ear mushrooms
1 onion
1 sweet red pepper
1 sweet green pepper
ginger to taste
15ml light soya sauce
5ml mature vinegar
3g sugar
10ml sesame oil
3g granulated chicken bouillon (or MSG)
3g salt

TIPS:

The dried wood ear mushrooms should be soaked in cold water. It is better to change the water several times to get a nice chewy texture.

The wood ear mushrooms should be cooked for 2-3 minutes after the water boils. If you soak the mushrooms in ice water afterward it will result in a better texture.

It will take longer to let flavour in wood ear mushrooms, so after sprinkling the seasoning sauce, wait for a while before eating and the dish will be tastier.

DIRECTIONS:

1. Peel the onion. Cut it in half and slice it.
2. Soak the dried wood ear mushrooms in cold water.
3. Julienne the ginger, red and green peppers for later use.
4. Heat water in a pot. Boil the soaked wood ear mushrooms untill cooked and transfer them into ice water.
5. Heat water in a pot. Boil the red and green peppers until cooked and transfer them into ice water.
6. Mix the light soya sauce, mature vinegar, sugar, sesame oil, granulated chicken bouillon and salt into a seasoning sauce.
7. Blend the onion, ginger, red and green pepper shreds and wood ear mushrooms together. Sprinkle the seasoning sauce and mix well. Allow them to sit for 30 minutes so as to let the flavour in. Serve immediately.

EASY CHINESE:

tiánjiāo
甜椒 sweet pepper

wèijīng
味精（鸡精）MSG (chicken bouillon)

yín'ěr
银耳 tremella

Xiānglàhǎidàisī
香辣海带丝
Spicy Seaweed Strips

INGREDIENTS:

200g seaweed strips

ginger, garlic, scallion, and small red peppers to taste

10ml light soya sauce

5ml mature vinegar

10ml sesame oil

3g granulated chicken bouillon (or MSG)

3g salt

TIPS:

Seaweed strips are coated with mucus and have the fishy smell peculiar to seafood, so boiling them will make them taste better.

Chinese people often like sour and spicy seaweed strips, because the seaweed will be crispier.

Adding garlic and sesame oil to the dish will offset the fishy smell from the seaweed.

Stir-frying the garlic and red peppers in hot oil until fragrant will add a special flavour to the dish.

DIRECTIONS:

1. Rinse the seaweed strips and soak them in water for 30 minutes.

2. Boil water in a pot, cook the seaweed strips and transfer them into ice water.

3. Cut the ginger, garlic and scallion into tiny pieces, and slice the small red peppers along the width.

4. Mix the light soya sauce, mature vinegar, sesame oil, granulated chicken bouillon to make the seasoning sauce.

5. Heat oil in a wok. Put in the ginger, garlic and small red peppers and stir-fry until fragrant.

6. Put the stir-fried ginger, garlic, small red peppers and scallions in the seasoning sauce, and mix well.

7. Drain the seaweed strips and put them in a bowl. Sprinkle the seasoning sauce and mix. Serve immediately.

EASY CHINESE:

Nǐ néng chī là de ma?
你 能 吃辣的 吗 ?
Can you eat spicy food?

Wǒ néng chī yìdiǎnr.
我 能 吃一点儿。
I can eat a little.

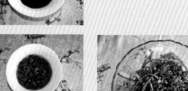

Tángcùfèngwěibáicài
糖醋凤尾白菜
Sweet and Sour Cabbage

INGREDIENTS:

1 Chinese cabbage
small red peppers and coriander to taste
5ml white vinegar
8g sugar
8g olive oil
3g salt

TIPS:

Cutting and slicing skills are required to cook this dish, so you need to have a sharp knife.

Since the stalks of cabbage are strong, we need to cut deeper without severing them.

The cabbage slices shouldn't be too thick.

When the cabbage is soaked in ice water, the slices will coil and make the dish more presentable.

The cabbage is sweet and fresh in itself, so only a small amount of sugar, vinegar sauce and olive oil is needed for this dish, which will make this dish taste and look better.

DIRECTIONS:

1. Peel the Chinese cabbage and divide its leaves and stalks.

2. Cut the stalks along their width at a 15 to 25 degree angle to the bottom without severing them while leaving a small distance between the cuts.

3. Julienne the chopped stalks vertically.

4. Soak the cabbage slices in ice water until they coil into circles.

5. Mix the white vinegar, sugar and salt into a seasoning sauce.

6. Blend the cabbage, coriander, small red peppers with olive oil. Sprinkle the seasoning sauce. Serve.

EASY CHINESE:

xiāngcài
香菜 coriander

gǎnlǎnyóu
橄榄油 olive oil

Zhège cài zhēn piàoliang!
这个菜真漂亮！
The dish looks so nice!

Shí shū dà liáng bàn
时蔬大凉拌
Cold Dressed Vegetables

INGREDIENTS:

half red cabbage
1 endive
12 cherry tomatoes
1 cucumber
10g olive oil
15ml apple vinegar
3g salt
1 piece of ginger

TIPS:

Olive oil and apple vinegar are the best choice for the dish.

Fruit vinegar, such as apple vinegar, will make the dish tastier and more nutritious.

When a mixture of olive oil, vinegar and ginger juice is added, the dish will become more delicious.

DIRECTIONS:

1. Rinse the cabbage and cut into small sections.

2. Remove the roots and yellow leaves from the endive, rinse and soak it in lightly salted water.

3. Cut the cherry tomatoes in half and slice the cucumber.

4. Mix the apple vinegar and olive oil into a seasoning sauce.

5. Squeeze the juice out of the ginger and blend it with the seasoning sauce.

6. Mix the vegetables together, sprinkle the mixed sauce.

EASY CHINESE:

zǐ gānlán
紫甘蓝 red cabbage

shèngnǚguǒ
圣女果 cherry tomato

kǔjú
苦菊 endive

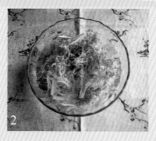

Xiǎocōngbàndòufu
小葱拌豆腐
Tofu Mixed with Chopped Scallions

INGREDIENTS:

1 tender tofu block
1 bundle of spring scallions
1 red sweet pepper
garlic to taste
15ml cooking oil
8g salt
5ml light soya sauce
3g granulated chicken bouillon (or MSG)

TIPS:

Tofu mixed with spring scallions is a delicious Chinese dish. It is better to use tender tofu for this dish.

Steam tender tofu for a while to make it healthier and tastier.

This dish can be great or it can be awful: the key is to bring out the fragrance and taste of the spring scallions. Firstly, make sure to fry the stalk in oil for its fragrance, and secondly, use the scallion leaves for colour and to adjust taste. The above two points are the secret of the dish.

DIRECTIONS:

1. Steam the tofu for 3 minutes.
2. Cut the tofu into any shape you would like after cooling.
3. Separate the stalk and leaves of the spring scallions. Cut the stalk into sections and chop the leaves. Chop the red sweet pepper. Mince the garlic.
4. Heat oil in a wok. Fry stalk over high heat in oil until fragrant. Remove stalk when it becomes yellow-brown and withered, then discard.
5. Stir-fry the chopped red sweet pepper and minced garlic in the oil until fragrant.
6. Add light soya sauce, salt and granulated chicken bouillon to the tofu block. Sprinkle the leaves of the scallions. Pour hot oil, chopped pepper and minced garlic on the tofu. Mix well and serve.

EASY CHINESE:

zhēng
蒸　to steam

zhǔ
煮　to boil

bàn
拌　to mix

Liángbànsānsī
凉拌三丝
Three Kinds of Mixed Vegetables

INGREDIENTS:

1 sheet of dried tofu
1 asparagus lettuce
1 carrot
Chinese prickly ash, ginger, garlic and dried pepper to taste
15ml light soya sauce
5g salt
3g granulated chicken bouillon (or MSG)

TIPS:

Three Kinds of Mixed Vegetables is a common home-style dish in China. Ingredients can be chosen and matched freely, such as shredded kelp, sliced white radish, fine Chinese vermicelli, sliced asparagus lettuce and carrot, etc. Every variation is delicious.

Fry the Chinese prickly ash in hot oil until fragrant to make Chinese prickly ash oil, which will add a special flavour to the dish.

Julienne vegetables as thin as possible for a better taste and presentation.

Colour is important in this dish, so vegetables of different colours should be used.

DIRECTIONS:

1. Rinse the sheet of dried tofu and julienne it.
2. Peel the carrot and asparagus lettuce and julienne them.
3. Chop ginger, mince garlic, seed and chop the dried pepper. Rinse Chinese prickly ash and dry it.
4. Boil water in a wok. Put in and cook the dried tofu, asparagus lettuce and carrot in succession.
5. Heat oil in a wok. Fry the Chinese prickly ash and then remove and discard.
6. Stir-fry the ginger, garlic and dried pepper in the Chinese prickly oil until fragrant. Season with light soya sauce, salt and granulated chicken bouillon and mix well.
7. Drain the dried tofu, asparagus lettuce and carrot and mix well. Sprinkle seasoning, mix well and serve.

EASY CHINESE:

Wǒ xǐhuan zuò liángbàncài.
我喜欢做凉拌菜。
I like cooking cold dressed dishes!

Zhè dào cài zuòfǎ hěn jiǎndān.
这道菜做法很简单。
This dish is very easy to cook.

Tángbànxīhóngshì
糖拌西红柿
Mixed Tomatoes with Sugar

INGREDIENTS:

4 tomatoes
50g sugar
honey to taste

DIRECTIONS:

1. Rinse the tomatoes and soak in lightly salty water.
2. Cut the tomatoes in half and remove the stem.
3. Place sliced tomatoes on a plate and sprinkle with honey.
4. Scatter sugar on the sliced tomatoes and then serve.

TIPS:

This is a fresh home-style cold dish in China, which, along with smashed cucumber, is the most common cold dish amongst Chinese families.

This cold dish tastes delicious when served immediately after scattering sugar, or after refrigerating for a while.

EASY CHINESE:

fēngmì
蜂蜜 honey

shātáng
砂糖 granulated sugar

bīngtáng
冰糖 crystal sugar

Bōcàibànfěnsī
菠菜拌粉丝
Spinach Mixed with Chinese Vermicelli

INGREDIENTS:

300g spinach
50g fine Chinese vermicelli
1 sweet red pepper
ginger to taste
15ml light soya sauce
10ml sesame oil
3g granulated chicken bouillon (or MSG)
5g salt

TIPS:

Make sure to boil the spinach first in order to separate the calcium oxalate out and also for better tasting spinach. This is also conducive to the prevention of calcium stones and assimilation of calcium.

Some vegetables are rich in chlorophyll. Adding a small amount of salt and cooking oil when cooking these vegetables in boiling water can make them look greener and more presentable.

The fine Chinese vermicelli should be soaked in cold water because this will make it more resilient and tastier.

DIRECTIONS:

1. Get rid of the yellow leaves and roots of the spinach. Rinse the spinach and soak it in lightly salted water for 10 minutes. Drain for later use.

2. Julienne the ginger. Seed and julienne the sweet red peppers.

3. Pour water into a pot, and then add a small amount of salt and a few drops of oil. Boil the water, then put the spinach in the boiling water and cook for 2 minutes.

4. Soak the hot spinach in ice water.

5. Soak the fine Chinese vermicelli in cold water while also boiling water. Put the soaked fine vermicelli in the boiling water and cook for 2 minutes, then immerse in ice water.

6. Boil water in a pot. Put in sweet red peppers and cook for 1 minute, then transfer them into ice water.

7. Mix the light soya sauce, sesame oil, granulated chicken bouillon and salt into a seasoning sauce. Mix and stir the spinach, sweet red peppers, ginger and fine Chinese vermicelli. Then sprinkle the sauce and mix again. Serve immediately.

EASY CHINESE:

yóucài
油菜　Chinese vegetable

shēngcài
生菜　lettuce

luóbo
萝卜　radish

Qiàngbànshēngcài
炝拌生菜
Lettuce Dressed with Hot Oil

INGREDIENTS:

300g lettuce

ginger, garlic, dried peppers and Chinese prickly ash to taste

15ml cooking oil

10ml light soya sauce

3g salt

3g granulated chicken bouillon (or MSG)

TIPS:

Heated oil with the fragrance of Chinese prickly ash will add a special flavour to the dish.

If dried peppers are added to a dish, their seeds should be removed for a better texture and presentation.

DIRECTIONS:

1. Rinse the lettuce and remove its roots. Soak it in lightly salted water.

2. Cut the ginger and garlic into tiny pieces. Cut the dried peppers into sections and remove their seeds. Clean and dry the Chinese prickly ash.

3. Heat oil in a wok. Stir-fry the Chinese prickly ash until fragrant, then remove the residue.

4. Put the ginger, garlic and dried peppers in the heated oil, then stir-fry them until fragrant.

5. Drain the lettuce, tear it into servable pieces and place in a container. Add the light soya sauce, salt and granulated chicken bouillon into the heated oil with the ginger, garlic and dried peppers in to make a seasoning sauce. Sprinkle the sauce and mix it with the lettuce well.

EASY CHINESE:

huājiāo
花椒 Chinese prickly ash

hújiāo
胡椒 pepper

gān làjiāo
干辣椒 dried pepper

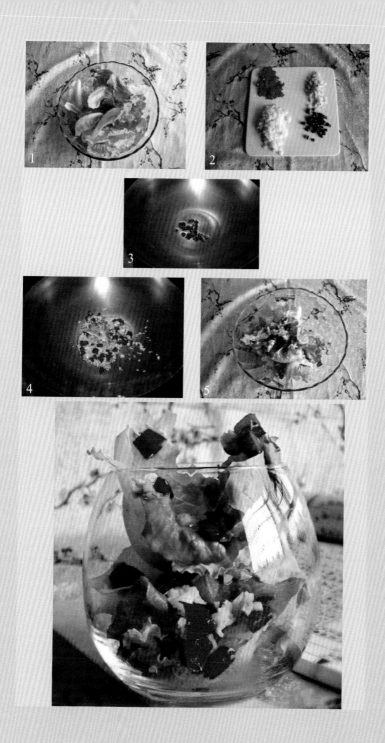

PART 2
热炒
Stir-frying

Jiāchángdòufu
家常豆腐
Home-style Tofu

INGREDIENTS:

1 block of tofu

10g dried wood ear mushrooms

1 sweet green pepper

1 sweet red pepper

garlic and ginger to taste

15ml light soya sauce

3g granulated chicken bouillon (or MSG); 5g salt

10ml sesame oil

100ml meat bouillon

TIPS:

Hard tofu should be used for frying instead of tender tofu. Also, tofu used for cooking should be steamed rather than braised because braised tofu will smell smoky.

Tofu should be put in cold storage because excess water will be pushed out and it will become more solid.

Chopped tofu should be rinsed off in water before cooking. If not, some small pieces will occur in cooking.

The wood ear mushrooms should be cooked before tofu with salt since the flavour won't penetrate into the wood ear mushrooms as easily as the tofu.

Add sesame oil before the dish is plated because the remaining heat will unlock the fragrance of the sesame oil and strengthen the fragrance and flavour of the dish.

DIRECTIONS:

1. Put the tofu in cold storage for 1 hour, and then drain of any excess water. Slice the tofu into thick pieces first then cut the slices diagonally. Rinse the tofu and remove the shreds.

2. Seed the green and red sweet peppers and chop them into diamond-shaped blocks. Cut the garlic and ginger into pieces, soak and clean the dried wood ear mushrooms.

3. Drain the tofu and fry it in the oil until the surfaces become golden.

4. Heat oil in a wok. Put in the garlic and ginger and stir-fry until fragrant. Put in the sweet red and green peppers and stir-fry for 1 minute.

5. Stir-fry the wood ear mushrooms and add salt.

6. Put in the fried tofu and stir-fry for 2 minutes.

7. Add light soya sauce and granulated chicken bouillon, then stir-fry until mixed well. Pour the meat bouillon in and cook until no liquid remains. Sprinkle sesame oil and turn off heat. Stir-fry again using the residual heat until mixed well.

EASY CHINESE:

Zhè kě shì wǒ de náshǒucài!
这可是我的拿手菜!
This dish is my specialty!

Zhēn xiāng a!
真香啊!
It is delicious!

Làchǎokōngxīncài
辣炒空心菜
Stir-fried Water Spinach with Pepper

INGREDIENTS:

400g water spinach

ginger, garlic and dried pepper to taste

5g salt

15ml light soya sauce

3g granulated chicken bouillon (or MSG)

TIPS:

Soak the green vegetables in the lightly salted water as this will help to remove the remaining pesticide and insects.

Since different times are used for cooking the stalks and leaves, they should be cooked successively.

Stalks should be cracked before cooking, because this will let in more flavour and reduce the cooking time.

Fresh leaves should be put in the boiling water with oil and salt for a while as this will make the leaves greener.

DIRECTIONS:

1. Rinse the water spinach and remove the withered leaves and stalks. Then divide the fresh leaves and stalks, and cut each stalk into several sections.

2. Soak the fresh leaves in lightly salted water for cleaning. Then boil water with a small amount of cooking oil and salt in it. Next immerse the fresh leaves into the boiling water for 30 seconds and then transfer them to ice water.

3. Julienne the ginger, cube the garlic, and section and seed the dried peppers.

4. Heat oil in a wok. Put in the ginger, garlic and dried peppers and stir-fry until fragrant.

5. Put in the stalks and add the salt. Stir-fry until the stalks become soft.

6. Put in the fresh leaves and sprinkle some light soya sauce and granulated chicken bouillon. Stir until mixed well, and serve.

EASY CHINESE:

Wǒ dì-yī cì chī zhège cài.
我 第一次吃 这个 菜。

This is my first time having this dish.

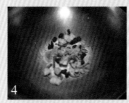

Jiàngbàoshūcàidīng
酱爆蔬菜丁
Sautéed Vegetable Dices in Sauce

INGREDIENTS:

1 cucumber

1 stem lettuce

12 dried Chinese black mushrooms

80g peanuts

1 sweet red pepper

spring onion and garlic to taste

2g salt

20g soybean paste

10ml light soya sauce

3g granulated chicken bouillon (or MSG)

10ml sesame oil

TIPS:

Since soybean paste and light soya sauce contain salt, less or no salt should be used.

Ingredients should be cooked in succession based on cooking time: Chinese black mushrooms first, then the cucumber and stem lettuce last.

Add the sesame oil before the dish is plated because the remaining heat will unlock the fragrance of the sesame oil and strengthen the fragrance and flavour of the dish.

DIRECTIONS:

1. Put a large amount of peanut oil in a wok and add the peanuts before the oil becomes hot. Fry them slowly until fragrant, then remove them from the oil allowing them to cool and dry.

2. Soak the dried Chinese black mushrooms and peel the stem lettuce, then rinse both of them; remove both ends of the cucumber and clean it, then cut it into small cubes.

3. Crack the garlic and cut it into sections. Then cut the sweet red pepper into sections and chop the spring onion.

4. Blend soybean paste, light soya sauce and granulated chicken bouillon together into the seasoning sauce.

5. Heat the oil and sauté the garlic and the sweet red pepper.

6. Put in the Chinese black mushrooms, cucumber and lettuce pieces successively, add the salt, and stir-fry them for 2 minutes.

7. Add the blended seasoning and stir-fry.

8. Scatter the spring onion and peanuts and turn off the heat, then sprinkle the sesame oil and stir-fry until mixed well. Then serve immediately.

EASY CHINESE:

qiē dīng
切 丁 to cube

qiē piàn
切 片 to slice

qiē sī
切 丝 to julienne

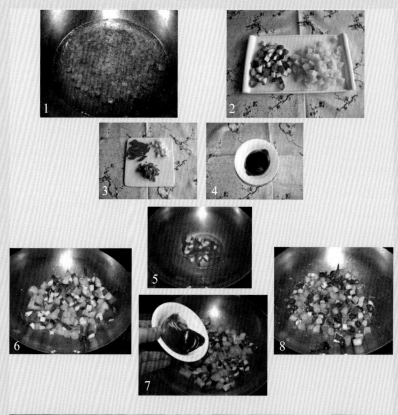

Cùliūbáicài
醋熘白菜
Stir-fried Chinese Cabbage with Vinegar

INGREDIENTS:

half Chinese cabbage

1 sweet red pepper

garlic and spring onion to taste

6g salt

10ml light soya sauce

5ml mature vinegar

3g granulated chicken bouillon (or MSG)

TIPS:

Since different cooking times are needed for the leaves and stalks of the Chinese cabbage, they should be cooked until soft in succession.

The stalks should be cut at an angle so that more inside area will be exposed. Due to this, more flavour will be let in and cooking time will be reduced.

Vinegar should be added twice because the vinegar added the first time will boost the absorption of the seasoning, the second time will add flavour to the dish.

DIRECTIONS:

1. Get rid of the old leaves and roots of the Chinese cabbage, then divide it into leaves and stalks.

2. Cut them into sections.

3. Cut the sweet red pepper into sections. Cut the garlic into small pieces.

4. Heat oil in a wok, then stir-fry the garlic and sweet red pepper until fragrant.

5. Put the stalks into the wok, pour in half of the prepared vinegar and all of the light soya sauce, then stir-fry the stalks until they let in flavour and become soft.

6. Put the leaves in the wok, add salt and another half of the vinegar. Stir-fry the leaves until they let in the flavour and become soft. Scatter some spring onion and stir-fry until fragrant, then serve immediately.

EASY CHINESE:

Zhège cài hěn shuǎngkǒu.
这 个 菜 很 爽 □。

This dish is very tasty.

Qīngchǎoxīlánhuā
清炒西兰花
Stir-fried Broccoli

INGREDIENTS:

1 head of broccoli
ginger and garlic to taste
1 sweet red pepper
3g salt
10ml light soya sauce
3ml mature vinegar
10ml oyster sauce

TIPS:

Soaking the broccoli in lightly salted water for a longer period will help to clean it more thoroughly because its flowers are closely clustered.

Stir-frying the broccoli with vinegar to taste will strengthen the flavour of the broccoli. Vinegar should be added in the end so that its flavour won't be lost.

Garlic matches with the stir-fried broccoli best. More garlic in the dish will make it tastier.

Adding some chopped sweet red peppers will make the dish look and taste better.

DIRECTIONS:

1. Get rid of the old root, and break the rest into pieces. Soak them in lightly salted water for 30 minutes then rinse. Cut the ginger into small cubes and chop the sweet red pepper. Cut the garlic into several sections.

2. Boil water with some salt and cooking oil in it. After the water is boiled, put the broccoli in and cook for 2 minutes, then soak it in ice water.

3. Heat oil in a wok and put the ginger and garlic in the oil, then stir-fry until fragrant.

4. Stir-fry the sweet red pepper until fragrant.

5. Put in the broccoli and add salt, then stir-fry for 1 minute.

6. Add the light soya sauce and oyster sauce, and stir-fry for 2 minutes. Sprinkle the mature vinegar and stir-fry again until there is a heavy vinegar fragrance. Serve immediately.

EASY CHINESE:

Wǎnfàn hěn fēngshèng!
晚饭很丰盛！
There was rich food for dinner!

Shuǐqíncàichǎoxiānggān
水芹菜炒香干
Stir-fried Celery with Smoked and Dried Tofu

INGREDIENTS:

- 250g smoked and dried tofu
- 100g celery
- 1 sweet red pepper
- ginger and garlic to taste
- 5g salt
- 10ml light soya sauce
- 10ml oyster sauce
- 3g granulated chicken bouillon (or MSG)
- 10ml sesame oil

TIPS:

Stir-frying the dried tofu until it become slightly dry in oil will give it a better texture.

The celery and dried tofu should be cooked successively since less time will be used for cooking the celery. Otherwise, the celery won't be crisp.

Fibres in celery are thick, so salt needs to be added first when it is cooked.

Add the sesame oil before the dish is plated because the remaining heat will unlock the fragrance of the sesame oil and strengthen the flavour of the dish.

DIRECTIONS:

1. Discard the celery leaves, clean the celery and slice it into pieces. Clean the smoked and dried tofu, cut it into big pieces at an angle, and rinse the pieces to remove the shreds on their surface.

2. Julienne the ginger and sweet pepper and slice the garlic.

3. Heat oil in a wok and put the tofu pieces in it. Stir-fry until their surfaces become slightly dry, then remove for later use.

4. Heat oil in the wok and add the ginger, garlic and red peppers. Stir-fry until fragrant.

5. Put the celery in with salt. Stir-fry for 1 minute.

6. Put the tofu in and stir-fry for 1 minute.

7. Add granulated chicken bouillon, light soya sauce and oyster sauce and stir-fry for 1 minute. Sprinkle the sesame oil, then turn off the heat and continue stir-frying until fragrant. Serve immediately.

EASY CHINESE:

Qíncài yǒuzhù xiāohuà.
芹菜有助 消化。
Celery helps digestion.

Xīhúluchǎojīdàn
西葫芦炒鸡蛋
Stir-fried Summer Squash with Eggs

INGREDIENTS:

3 eggs
1 summer squash
1 sweet red pepper
ginger and spring onion to taste
8g salt
10ml light soya sauce

TIPS:

A small amount of water and several drops of cooking wine should be added to the eggs before whisking as it will make the cooked eggs fluffier.

More oil, higher heat and quick beating will give the eggs a nice texture.

Excessive use of seasoning will mask the taste of the eggs.

DIRECTIONS:

1. Crack the eggs into a container. Add several drops of cooking wine and a spoon of warm water. Whisk the eggs quickly and add some salt.

2. Peel the summer squash and clean it. Then cut it into several sections.

3. Cut the ginger, sweet pepper and spring onion to tiny pieces.

4. Put a larger amount of oil in a wok. After the oil is heated up, put the eggs in and stir-fry until they become fluffy. Allow the oil to drain off the eggs.

5. Heat oil in the wok and put the ginger and sweet pepper in and stir-fry until fragrant.

6. Put in the summer squash with salt and some light soya sauce. Stir-fry for 2 minutes.

7. Add the cooked eggs, sprinkle the rest light soya sauce and stir-fry them well. Scatter some spring onion and stir-fry until fragrant.

EASY CHINESE:

Zhège cài hěn shìhé xiǎopéngyǒu.
这个菜很适合小朋友。
This dish is suitable for children.

Jiàngshāoqiézi
酱烧茄子
Sauteed Eggplant with Soybean Paste

INGREDIENTS:

3 eggplants
ginger, garlic and spring onion to taste
1 sweet green pepper
1 sweet red pepper
2g salt; 20g soybean paste
5ml light soya sauce
10ml oyster sauce
3g granulated chicken bouillon (MSG)
10ml sesame oil

TIPS:

Since soybean paste and light soya sauce contain salt, little or no salt should be used.

Eggplants easily oxidize, so they should be soaked in lightly salted water to avoid a colour change.

Salt is easily absorbed by eggplant, so seasoning containing salt should be used sparingly.

Hot oil and quick beating are the keys to cooking eggplants.

Add the sesame oil before the dish is plated because the remaining heat will unlock the fragrance of the sesame oil and strengthen the flavour of the dish.

DIRECTIONS:

1. Clean the sweet peppers and cut them into sections. Quarter the ginger, chop the garlic and cut the spring onion into sections.

2. Clean the eggplants and remove both ends. Cut into several sections and soak in lightly salted water to keep fresh.

3. Heat oil in a wok, and stir-fry the ginger and garlic until fragrant.

4. Add the sweet pepper and salt, and stir-fry until the pepper becomes a little soft.

5. Put the blend of the soybean paste, light soya sauce, oyster sauce, granulated chicken bouillon and a small amount of water, and then bring it to a boil on high heat.

6. While the seasoning sauce is boiling, another wok should be heated with oil. Fry the eggplants for 30 seconds and then remove and drain them of oil.

7. Put the fried eggplants into the wok with the seasoning sauce. Stir-fry on high heat so that the seasoning sauce will coat all of the eggplant and more flavour will penetrate them. Sprinkle some spring onions and sesame oil when the eggplant flesh becomes soft and changes colour. Then turn off the heat and stir-fry until fragrant.

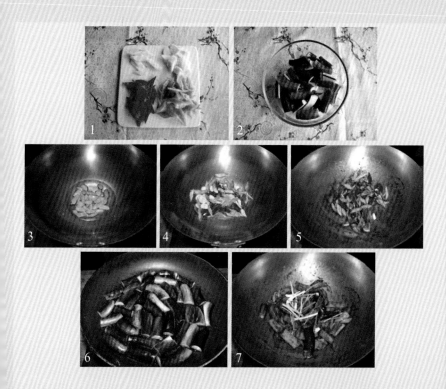

Hóngtángmìlián'ǒu
红糖蜜莲藕
Preserved Lotus Root with Brown Sugar

INGREDIENTS:

2 lotus roots
80g brown sugar
white vinegar

TIPS:

Lotus roots contain starch, so rinse off the starch after they are cut for a better texture.

The selected brown sugar should be cube-shaped with clear layers, because this kind of sugar is of high quality. In addition to this, black sugar is also a good choice.

Cook the lotus root before adding brown sugar. In this way, the starch on the root surface will be removed to get a better chewy texture.

Remove the lotus root when there is 10% brown sugar juice left. Cook the juice separately until condensed to avoid the lotus root sticking to the pot.

Sprinkle some white sesame on the lotus root to add flavour and colour to the dish.

DIRECTIONS:

1. Clean the lotus roots, remove both ends and peel them.

2. Slice the lotus roots thin, place them in a bowl and rinse off the starch on their surface.

3. Crush the brown sugar cubes into powder.

4. Heat water in a pot, and add a small amount of white vinegar after the water has been boiled. Put the lotus roots in the pot and cook for 2 minutes. Remove them up and allow the water to drain off, then pour the water out of the pot.

5. Add fresh water to the pot. Put in the lotus root slices and brown sugar and cook them for 20 minutes.

6. Cook until 10% of the water is left in the pot. When the lotus roots reach a deep colour and absorb the flavour, place them in a bowl.

7. Cook the remaining sugar juice in the wok until thick and spread on the lotus root. Serve after the dish cools.

EASY CHINESE:

Tiántián de zhēn hǎochī!
甜 甜 的 真 好 吃！
It's sweet and delicious!

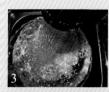

Xiānglà'ǒudīng
香辣藕丁
Spicy Lotus Root Cubes

INGREDIENTS:

1 lotus root

ginger, garlic, spring onion and dried pepper to taste

5g salt

10ml light soya sauce

10ml oyster sauce

3g granulated chicken bouillon (or MSG)

10ml sesame oil

3ml white vinegar

TIPS:

The lotus roots contain starch, so rinse off the starch after they are cut for a better texture.

Lotus root cubes are thick, so boil them first so they will cook more easily.

Remove the seeds after the dried peppers are mashed. This will give the dish a nice flavour and presentation.

Sprinkle white vinegar while cooking the lotus root cubes to prevent the cubes from oxidizing and becoming black.

DIRECTIONS:

1. Peel the lotus root and cut into cubes. Rinse the starch off on the surface.

2. Heat water in a pot and add a small amount of white vinegar after the water boils. Put the lotus root cubes in the pot and cook for 2 minutes. Remove and allow water to drain off.

3. Cut the ginger, garlic and spring onion into pieces. Mash the dried pepper and remove its seeds.

4. Heat oil in a wok. Add the ginger and garlic and cook until fragrant.

5. Add the lotus root cubes and white vinegar, and stir-fry for 1 minute.

6. Add the light soya sauce, oyster sauce, salt and granulated chicken bouillon and scatter dried pepper. Stir-fry for 2 minutes.

7. Scatter spring onion and turn off the heat. Sprinkle the sesame oil and stir-fry until mixed well and fragrant. Serve immediately.

EASY CHINESE:

Tài là le!
太辣了!
It's too spicy!

Hē diǎnr chéngzhī ba.
喝点儿 橙 汁 吧。
Have some orange juice, please.

Tángcùqīngjiāo
糖醋青椒
Sweet and Sour Green Peppers

INGREDIENTS:

300g green peppers

ginger, garlic and spring onion to taste

5g white sugar

8ml mature vinegar

15ml light soya sauce

5g salt

3g granulated chicken bouillon (or MSG)

TIPS:

Remove the white flesh and seeds inside the peppers to get a nicer texture.

Adding more vinegar into the seasoning sauce will make dishes with pepper more delicious.

Stir-fry the peppers without the oil so that their skin will dehydrate and the peppers will become more fragrant and have a better texture.

Sprinkle the seasoning sauce and stir-fry the peppers on high heat to let in more flavour.

DIRECTIONS:

1. Clean the peppers and remove the ends. Crack or cut them open and get rid of the white flesh and seeds inside, then cut them into slices.

2. Cut the ginger, garlic and spring onion into pieces.

3. Mix the white sugar, mature vinegar, light soya sauce and granulated chicken bouillon into a seasoning sauce.

4. Stir-fry the pepper slices without oil until their skin becomes dry, and then remove them for later use.

5. Heat oil in a wok. Put in ginger and garlic and stir-fry until fragrant. Put in pepper slices and salt and stir-fry for 1 minute.

6. Sprinkle the seasoning sauce and stir-fry for 2 minutes until no liquid is left.

7. Scatter the spring onion and stir-fry until mixed well.

Hétángxiǎocài
荷塘小菜
Stir-fried Lotus Root and Snow Peas

INGREDIENTS:

150g snow peas
1/2 lotus root
1 carrot
10g dried wood ear mushrooms
1 sweet red pepper
garlic and spring onion to taste
8g salt
15ml light soya sauce
3g granulated chicken bouillon (or MSG)
10ml sesame oil

TIPS:

Remove the strings of snow pea for a better texture.

Raw peas are inedible, so they should be boiled in water before cooking. If cooking oil and salt are added to the water, the peas will become greener.

Different times will be spent on cooking different ingredients, so the ingredients should be cooked in succession to maintain their colours and texture.

DIRECTIONS:

1. Rinse all the vegetables. Remove the ends and strings of the snow peas, then cut them into sections. Peel and section the carrot and then slice the sections lengthwise. Soak the dried wood ear mushrooms and rinse. Peel and slice the lotus root.

2. Cut the sweet red peppers into slices, garlic into pieces and spring onion into thin sections.

3. Heat water in a pot with a small amount of salt and cooking oil and boil the snow peas, wood ear mushrooms and lotus root successively in the water.

4. Heat oil in a wok and stir-fry the garlic and sweet red peppers until fragrant.

5. Add the carrots and wood ear mushrooms and stir-fry for 1 minute.

6. Add the lotus root, snow peas, salt, light soya sauce and granulated chicken bouillon and then stir-fry for 2 minutes.

7. Scatter the spring onion, turn off the heat, sprinkle the sesame oil and stir-fry until mixed well.

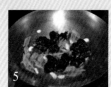

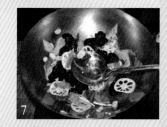

PART 3
小蒸
Steaming

Suànní qiézi
蒜泥茄子
Steamed Eggplant with Garlic

INGREDIENTS:

350g eggplants

ginger, garlic, scallion and red pepper to taste

5g salt

15ml light soya sauce

10ml sesame oil

3g granulated chicken bouillon (or MSG)

TIPS:

Peeled eggplants easily oxidize and their colour changes, so they should be soaked in lightly salted water to insulate themselves from the air.

Steam the whole eggplants and then tear them into strips instead of cutting them first and steaming to get a nicer chewy texture.

Mix the sesame oil with the eggplant strips immediately so as to insulate them and avoid a change of colour.

DIRECTIONS:

1. Remove the ends of the eggplants, peel and soak them in lightly salted water.

2. Mash the garlic, cut the spring onion and ginger into tiny pieces and the red pepper along the width.

3. Put the whole eggplants into a steamer and steam for 10 minutes.

4. Cool the steamed eggplants and tear them into strips. Put them in a bowl, sprinkle with sesame oil and mix well.

5. Heat oil in a wok. Put in the garlic, ginger and red pepper and stir-fry them until fragrant. Add the light soya sauce, granulated chicken bouillon and mix well.

6. Scatter the spring onion and sprinkle the blend on the eggplants. Mix well and serve.

EASY CHINESE:

Zěnme tiāoxuǎn qiézi?
怎么挑选茄子？
How to choose the eggplant?

qiézi duōshao qián yì jīn?
茄子多少钱一斤？
How much is the eggplant for a half kilo?

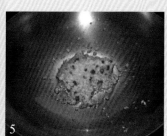

Duòjiāozhēngjīnzhēngū
剁椒蒸金针菇
Steamed Golden Needle Mushrooms with Chopped Peppers

INGREDIENTS:

200g golden needle mushrooms

chopped peppers and ginger to taste

15ml light soya sauce

3g salt

3g granulated chicken bouillon

8ml sesame oil

TIPS:

Since golden needle mushrooms grow closely, soil and dirt are easily hidden between their stalks, so soak them in lightly salted water for cleaning after getting rid of the roots and tearing them apart.

Chopped peppers often contain a lot of salty water, so drain the peppers before cooking.

Stir-fry the chopped peppers in hot oil until fragrant to get a nicer texture.

DIRECTIONS:

1. Rinse the golden needle mushrooms. Get rid of their roots and tear them apart. Soak them in lightly salted water for cleaning.

2. Cut the ginger into small pieces and drain the chopped peppers.

3. Heat oil in a wok, put in the ginger and pepper and stir-fry until fragrant.

4. Mix light soya sauce, salt and granulated chicken bouillon.

5. Drain the golden needle mushrooms and place them on a plate. Sprinkle the seasoning sauce on the golden needle mushrooms and steam them for 8 minutes. Take them out and sprinkle sesame oil. Serve immediately.

EASY CHINESE:

qǐng duō chī yìdiǎnr.
请 多 吃 一点儿。
Please have more.

Bú yào kèqi.
不 要 客气。
Make yourself at home.

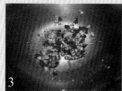

Nǎixiāngyùmǐxiǎo
奶香玉米小
nánguā
南瓜
Steamed Pumpkin and Creamy Corn Kernels

INGREDIENTS:

1 small pumpkin
50g corn kernels
10g sugar
1 spoon light cream
honey to taste

TIPS:

Use a small green pumpkin when cooking dessert, as green pumpkins are sweeter and fresher than the orange ones.

Chop half of corn kernels to release the fragrance of the corn.

Honey has a unique flavour which will add a special highlight to the dish.

Honey contains a large amount of active ingredients. The high temperature can kill them and make honey taste sour, so make sure to drizzle honey on the pumpkin after it cools down and the temperature reaches below 45 degrees Celsius.

DIRECTIONS:

1. Clean the small pumpkin. Remove the top end.
2. Use the spoon to remove the core. Carve the pumpkin out like a bowl.
3. Wash the corn kernels and drain them. Chop half of corn kernels. Leave the other half as is.
4. Mix the corn, sugar and light cream.
5. Put the mixture in the pumpkin "bowl". Place the pumpkin in the steamer over boiling water and steam for 20 minutes. Remove the pumpkin and let it cool down until the temperature drops below 45 degrees Celsius. Then sprinkle honey on the pumpkin and serve.

EASY CHINESE:

zhǔshí
主食 staple food

càiyáo
菜肴 dish

tiáopǐn
甜品 dessert

Gòngmǐniàng'ǒu
贡米酿藕
Stuffed Lotus Root with Purple Rice

INGREDIENTS:

1 lotus root
50g purple rice
30g sugar
honey to taste

TIPS:

Peeled lotus root should be soaked in water with white vinegar to stop oxidation and discolouration if it is not cooked immediately.

Purple rice is nutritious and has a hard texture. Make sure to soak it completely for a long time before steaming.

Make sure to stuff the purple rice in the lotus root effectively, so the dish looks good.

Honey has a unique flavour which will add a special highlight to the dish.

Honey contains a large amount of active ingredients. The high temperature can kill them and make honey taste sour, so make sure to drizzle honey on the pumpkin after it cools down and the temperature reaches below 45 degrees Celsius.

Osmanthus sugar can also be sprinkled on the lotus root to give the dish the fragrance of osmanthus flowers.

DIRECTIONS:

1. Wash the purple rice once, and then soak it in cold water for more than 12 hours.
2. Peel the lotus root. Remove both ends.
3. Remove the purple rice from water, drain it and mix it with the sugar well.
4. Cut one end of the lotus root 2cm from the end.
5. Stuff the purple rice into the lotus root. Use chopsticks to push the rice into the lotus root to effectively stuff it.
6. Use toothpicks or other tools to fasten the cut section in step 4 to the lotus root.
7. Put the whole lotus root in the steamer. Steam it over boiling water for 40 to 50 minutes. Then leave the lotus root to cool down. When the temperature reaches below 45 degrees Celsius, slice the lotus root, sprinkle honey and serve.

EASY CHINESE:

zǐmǐ
紫米 purple rice

nuòmǐ
糯米 glutinous rice

xiǎomǐ
小米 millet

Suànróngzhēngsīguā
蒜蓉蒸丝瓜
Steamed Towel Gourd with Mashed Garlic

INGREDIENTS:

2 towel gourds

garlic, scallions and sweet red peppers to taste

3g salt

10ml soya sauce

3g granulated chicken bouillon (or MSG)

5ml oyster sauce

TIPS:

Oxidation may take place when the towel gourd is exposed to air, which will cause discolouration, so the peeled towel gourds should be soaked in lightly salted water to stop the reaction.

Let the garlic cook in the oil until it turns yellow-golden in colour. The fragrance of the garlic will be released completely but the garlic itself will lose its flavour; this is why the remaining half of the garlic should be lightly cooked and then added to the dish again before serving.

A towel gourd has a gentle texture and can't bear a long time cooking. However, flavour does not easily sink in the sectioned towel gourd, so use toothpicks to stick through each section so as to let in more flavour.

DIRECTIONS:

1. Remove both ends of the towel gourds. Peel them and cut into 3cm sections. Thoroughly stick the sections using toothpicks or other tools to let in the flavour.

2. Mash the garlic. Chop the sweet red peppers and scallions.

3. Heat the oil in the wok. Add half of minced garlic and stir-fry slowly until fragrant and when it turns yellow-golden, and then remove it.

4. Place the towel gourd sections on the plate. Sprinkle the yellow-golden garlic on top of each section. Mix the fried garlic oil, salt, soya sauce, oyster sauce and granulated chicken bouillon and then sprinkle the mixture on the gourd sections. Boil the steamer on high heat. Steam the towel gourd for 6 minutes.

5. Remove the plate after steaming. Discard the yellow-golden garlic. Pour the liquid off of the plate and set aside.

6. Heat the oil in the wok. Stir-fry the rest of minced garlic, the chopped sweet red peppers and the chopped scallions and cook until fragrant. Add the remaining liquid and bring it to a boil. Then sprinkle them on the towel gourd sections and serve.

Bīngtángzhēngshānyào
冰糖蒸山药
Steamed Yam with Rock Sugar

INGREDIENTS:

1 Chinese yam

80g rock sugar

osmanthus flowers and cranberry to taste

TIPS:

Yam should be soaked in water to prevent oxidation and discoloration.

Peel the yam under running water to prevent oxidation and discolouration and begin to wash some of the sticky mucus off.

Oxidation can be completely stopped in water with white vinegar.

Yam should be boiled for a while in water with white vinegar to prevent oxidation while steaming.

It is better to use brown polycrystalline rock sugar because white polycrystalline rock sugar tastes sweet but has less nutritional value.

Sprinkle osmanthus flowers and chopped fruit when finished cooking to add pleasant taste combinations. Besides the chopped cranberry, other fruits can be added as you like.

DIRECTIONS:

1. Rinse the Chinese yam well, peel it under running water, cut off both ends and then soak it in the ice water with white vinegar.

2. Cut the yam into 6cm long sections. Make sure to wash the mucus off of the yam in the ice water with white vinegar.

3. Boil water in a pot. Add white vinegar to taste after the water boils. Add the yams and boil for 3 minutes.

4. Mash the rock sugar.

5. Wash the mucus off of the yam again and drain after. Place them on a plate. Sprinkle the sugar and steam over boiling water for 20 minutes.

6. Take the yam sections from the steamer. Pour the syrup exuded from sugar in a pot and bring it to boil. Then sprinkle the syrup, chopped osmanthus flowers and cranberry on the yam and serve.

EASY CHINESE:

zǐshǔ
紫薯 purple sweet potato

mànyuèméi
蔓越莓 cranberry

lánméi
蓝莓 blueberry

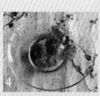

Duòjiāozhēngwáwacài
剁椒蒸娃娃菜
Steamed Baby Chinese Cabbage with Diced Hot Red Pepper

INGREDIENTS:

2 baby Chinese cabbages

diced hot red pepper, garlic and scallions to taste

10ml light soya sauce

3g granulated chicken bouillon (or MSG)

5g salt

TIPS:

Diced hot pepper has a large amount of salty water. Make sure to drain it before cooking.

Stir-fry diced hot peppers with oil until fragrant, so it will taste better.

Keep baby Chinese cabbage for a while in boiling water as this will remove its peculiar earthy smell.

DIRECTIONS:

1. Remove old leaves and the roots of the baby Chinese cabbages. Clean them. Mash the garlic. Chop the scallions. Rinse and drain the hot red pepper.

2. Cut the baby Chinese cabbage horizontally into 4-6 pieces.

3. Boil water in a pot. Keep them in boiling water for 2 minutes.

4. Heat oil in a wok. Stir-fry the mashed garlic and diced hot pepper until fragrant.

5. Add light soya sauce, salt and granulated chicken bouillon. Mix well.

6. Place the baby Chinese cabbage on the plate. Sprinkle the mixture on them. Steam them over boiling water for 8 minutes. Remove the plate and sprinkle chopped scallion. Spread hot oil, so as to make the dish tastier. Then serve.

EASY CHINESE:

Xué zuò Zhōngguócài hěn yǒu yìsi.
学 做 中国菜 很 有 意思。

It's very interesting to learn how to cook Chinese dishes.

Hóngzǎoniàngkǔguā
红枣酿苦瓜
Stuffed Bitter Melon with Jujubes

INGREDIENTS:

1 bitter melon
30 jujubes
30g sugar
osmanthus flowers to taste

TIPS:

The core of bitter melon is the main source of the bitter taste. Remove the seeds and white pith so that the bitter taste will be lessened.

Make sure to stuff jujubes effectively so that the dish looks good after slicing.

Besides water, other juices you prefer can also be used to boil syrup.

Slice the bitter melon as thin as possible so that it will taste and look better.

Ice the dish in the fridge after cooking to add more flavour to the dish.

DIRECTIONS:

1. Wash the bitter melon. Remove both ends. Cut it into 8-centimetre sections.
2. Remove the core using a spoon.
3. Rinse jujubes. Cut them vertically and remove the seeds.
4. Stuff the jujubes into the bitter melon. Make sure to stuff it effectively.
5. Place the melon in a steamer over boiling water and steam for 6 minutes. Remove and slice it.
6. Bring a small amount of water to boil in a wok. Add sugar. Boil it slowly into syrup. Then sprinkle the syrup and osmanthus flowers on the sliced bitter melon and serve.

EASY CHINESE:

sīguā
丝瓜 towel gourd

nánguā
南瓜 pumpkin

dōngguā
冬瓜 wax gourd

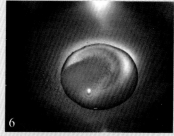

PART 4
微 煮
Boiling

Shuǐzhǔyùmǐ
水煮玉米
Boiled Corn

INGREDIENTS:

2 ears of corns
some salt and baking soda

TIPS:

If the corn is fresh and the husks remain clean and whole, they can be boiled with the corn to improve the flavour.

Add a little salt in the water to make the corn taste fresher and sweeter.

The main nutrient in corn is niacin. Add some jequirity beans to release more niacin and avoid the loss of vitamin B1 and B2 in the corn. However, only a small amount of beans can be added otherwise the dish will taste astringent.

DIRECTIONS:

1. Husk the corn and discard the husks and tassels.
2. Rinse the corns and cut into sections.
3. Boil water in a pot. Add a pinch of salt and baking soda.
4. Boil the corn sections for 20 minutes. Remove and cool to a proper temperature, then serve.

EASY CHINESE:

Wǎnshang wǒmen chī yùmǐ ba.
晚上 我们 吃 玉米 吧。
Let's eat corn tonight.

Qǐng bǎ yùmǐ qiēchéng duàn.
请 把 玉米 切成 段。
Please cut the corn into several segments.

Yánshuǐzhǔmáodòu
盐水煮毛豆
Boiled Green Beans in Salty Water

INGREDIENTS:

300g green beans

anise, cinnamon, dried pepper and ginger to taste

20g salt

20ml light soya sauce

5g granulated chicken bouillon (or MSG)

TIPS:

Green beans have small thorns, so scrubbing them using salt can get rid of the thorns and other impurity that we can't see clearly.

Remove both ends of the green beans so as to let in more flavour during cooking.

Boil green beans in heavily salted water. Use more salt than when cooking soups.

Do not use an iron pot to boil green beans, use glass containers instead, though earthenware is best.

Do not cover the pot when boiling green beans otherwise they will turn brown.

Soak the green beans in salty water after boiling to let in more flavour.

Use more seasonings if you can find them: such as, Chinese prickly ash, white pepper, myrcia, fennel and clove.

DIRECTIONS:

1. Rinse the green beans. Remove both ends and drain. Toss the green beans with a lot of salt in a container and then rinse them under running water.

2. Heat oil in a wok. Stir-fry the anise, cinnamon, dry pepper and ginger until fragrant.

3. Pour water in the wok.

4. Pour the mixture into an earthenware pot. Bring the mixture to boil on high heat. Season it with salt, light soya sauce, oyster sauce and granulated chicken bouillon.

5. Place the green beans into the earthenware pot. Bring them to a boil on high heat and continue to boil on low heat for 12 minutes. Then turn off the gas.

6. Soak the green beans in the salty water for 2 hours. Then remove and serve.

EASY CHINESE:

wāndòu
豌豆   pea

hélándòu
荷兰豆 snow pea

sìjìdòu
四季豆 green bean

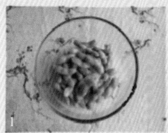
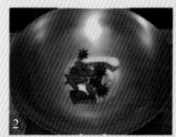
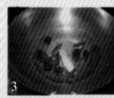

Wǔxiāngpíjiǔhuāshēng
五香啤酒花生
Five-flavour Peanuts in Beer

INGREDIENTS:

300g peanuts in shell

anise, cinnamon, myrcia and dried pepper to taste

20g salt

20ml light soya sauce

5g granulated chicken bouillon (or MSG)

TIPS:

Soaking unshelled peanuts is the first step, which can make the shells extend. It's the preparation for the next step: rinsing.

The shells of peanuts are not even, so they are not easy to clean. Use flour to scrub the shells so that hidden dirt and impurities can be removed completely.

Pinch the shell open before boiling, so as to let in more flavour.

Soak the boiled peanuts in the beer soup so as to let in more flavour.

DIRECTIONS:

1. Rinse unshelled peanuts under running water. Soak in water for 15 minutes and drain.

2. Scatter some flour on the peanuts. Scrub them several times and then rinse.

3. Pinch the shell open.

4. Rinse the seasoning.

5. Pour beer in an earthenware pot. Add seasoning. Bring it to boil on high heat. Season with salt, light soya sauce and granulated chicken bouillon.

6. Put peanuts in the pot. Bring them to boil on high heat and continue to boil on low heat for 20 minutes. Then turn off gas. Soak peanuts in the beer soup for 2 hours and then serve.

EASY CHINESE:

báijiǔ
白酒 liquor

hóngjiǔ
红酒 wine

huángjiǔ
黄酒 millet wine

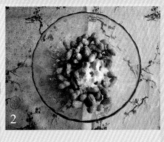

Shāguōbáicàidòufu
砂锅白菜豆腐
Simmered Chinese Cabbage and Tofu in an Earthenware Pot

INGREDIENTS:

1/2 tofu block

1/2 head of Chinese cabbage

1000ml soup broth (meat or bone)

ginger, scallion and Chinese wolfberry to taste

8g salt

15ml light soya sauce

3g granulated chicken bouillon (or MSG)

TIPS:

Cooking in an earthenware pot adds a unique flavour.

Boil tofu in salty water before cooking to prevent it from breaking apart during cooking.

Cooking time for cabbage leaves is shortest, so add them last. Bring the leaves to a boil and then remove.

Meat or bone soup used as a soup base when cooking vegetable soup will add much flavour to the dish.

DIRECTIONS:

1. Cube the tofu, then rinse it under running water.

2. Remove the root of the Chinese cabbage. Separate the leaves and stalks. Cut stalks at an angle into slices. Cut leaves into properly sized slices.

3. Julienne the ginger, chop the spring onion, soak the Chinese wolfberry.

4. Boil water in a pot. Add salt. Boil tofu in the water for 2 minutes.

5. Heat oil in the wok, put in ginger and stir-fry until fragrant.

6. Put in the stalks, stir-fry until stalks become soft. Add soup and bring to a boil.

7. Put in tofu, season with salt, light soya sauce and granulated chicken bouillon. Bring to a boil on high heat and then simmer on low heat for 20 minutes.

8. Put in leaves, boil them for 2 minutes. Scatter Chinese wolfberries and boil for 1 minute. And then sprinkle spring onion and remove.

EASY CHINESE:

Dòufu hányǒu hěn duō dànbáizhì.
豆腐 含有 很 多 蛋白质。
Tofu contains a lot of protein.

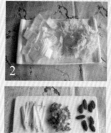

Jīnzhēngūdòufutāng
金针菇豆腐汤
Golden Needle Mushroom and Tofu Soup

INGREDIENTS:

half tofu block

a bundle of golden needle mushroom

ginger, scallion and Chinese wolfberry to taste

1000ml soup broth (meat or bone)

8g salt

10ml light soya sauce

3g granulated chicken bouillon (or MSG)

TIPS:

Boil tofu in salty water before cooking to prevent it from breaking apart during cooking.

Golden needle mushrooms have thick stem, so dirt and other substances are easily hidden among the stalks. Make sure to tear the stalks up and rinse them in lightly salty water.

When cooking mushroom soup, the mushrooms should be stir-fried in heated oil before boiling. It will add flavour.

Meat or bone soup used as soup base when cooking vegetable soup will add much flavour to the dish.

DIRECTIONS:

1. Cube the tofu block and rinse it under running water.

2. Rinse the golden needle mushrooms under running water, remove roots and tear stalks into pieces and soak in lightly salty water to clean.

3. Boil water in a pot, add salt. Boil tofu in the water for 2 minutes.

4. Heat oil in a wok. Stir-fry the ginger until fragrant.

5. Put in the golden needle mushroom and stir-fry for 1 minute, pour in the soup and bring it to a boil.

6. Put in tofu, season with salt, light soya sauce and granulated chicken bouillon. Bring it to a boil on high heat and then simmer on low heat for 20 minutes. Scatter Chinese wolfberries and boil for 1 minute. Then sprinkle scallion, remove from heat and serve.

EASY CHINESE:

xìngbàogū
杏鲍菇 trumpet mushroom

jītuǐgū
鸡腿菇 ink cap mushroom

xiānggū
香菇 fragrant mushroom

Gūxiāndòufu
菇鲜豆腐
Mushroom and Tofu

INGREDIENTS:

500g various kinds of mushrooms

ginger, scallion, dried pepper, fermented soya beans and Chinese prickly ash to taste

5g salt

15ml light soya sauce

10ml oyster sauce

1000ml soup broth (meat or bone)

ground pepper to taste

TIPS:

Boil tofu in salty water before cooking to prevent it from breaking apart during cooking.

When cooking mushroom soup, the mushrooms should be stir-fried in hot oil before boiling. It will add flavour.

Braised dishes such as this one all have a heavy taste. So using meat or bone soup broth as a base will make the dish tastier.

When braising vegetables, seasoning with Chinese prickly ash and ground pepper can add a lot of flavour.

Fermented soya beans are a fantastic seasoning, which can be used for stir-frying and boiling. It will improve the flavour and make the dish taste fresher.

DIRECTIONS:

1. Cube the tofu block and rinse it under running water.

2. Rinse all mushrooms under running water. Put the mushroom umbrellas face down and soak in lightly salty water for 15 minutes. Slice or cube the mushrooms.

3. Slice the ginger, cut the scallion into sections and rinse the dried pepper and Chinese prickly ash.

4. Boil water in a pot. Add salt and boil tofu in the water for 2 minutes.

5. Heat oil in a wok. Stir-fry ginger, dried pepper, fermented soya beans and Chinese prickly ash until fragrant.

6. Add mushrooms and stir-fry for 2 minutes.

7. Add the soup broth. Bring them to a boil on high heat and simmer for 10 minutes on low heat.

8. Add tofu and simmer for 20 minutes. Season it with salt, light soya sauce and oyster sauce and continue to simmer for 10 minutes.

9. Sprinkle ground pepper and scallions. Allow the sauce to boil off over high heat and remove the wok.

Shūcàibǎojiàntāng
蔬菜保健汤
Vegetable Soup

INGREDIENTS:

1 carrot
1 ear of corn
20g dried Chinese shitake mushrooms
2 tomatoes
15g salt
10ml light soya sauce
1000ml soup broth (meat or bone)

TIPS:

Since different cooking and simmering times are needed for different vegetables, they should be cooked in succession in order to bring out the best flavour and texture.

Meat or bone soup broth used as soup base when cooking vegetable soups will make the dish tastier.

You can add other ingredients you like for this soup; Chinese yams, all kinds of mushrooms and other vegetables also can be added.

DIRECTIONS:

1. Soak the dried Chinese shitake mushrooms in lukewarm water.

2. Clean the corn, carrot and tomatoes and peel the carrot. Then chop the carrot and tomatoes into cubes.

3. Sauté the sliced ginger until fragrant and then pour the soup broth and ginger into an earthenware pot, then add the mushrooms. Simmer for 20 minutes.

4. Add the corn. Simmer for another 20 minutes.

5. Add the carrot, salt and soya sauce. Simmer for another 20 minutes.

6. Add the tomatoes. Boil for 2 minutes. Then remove the pot from the heat and serve.

EASY CHINESE:

Tài hǎo hē le!
太 好 喝 了！
The soup is delicious!

Wǒ yào zài hē yì wǎn.
我 要 再 喝一碗。
I want one more bowl.

策　　划：付　眉　阮　芳
责任编辑：韩　　颖
英文编辑：韩芙芸
封面设计：新　　乐

图书在版编目（CIP）数据

　　我爱中国菜·素食篇：英文 / 心蓝编著 .—北京：华语教学出版社，2015.1
　　ISBN 978-7-5138-0929-0

　　Ⅰ. ①我… Ⅱ. ①心… Ⅲ. ①素菜—菜谱—中国—英文 Ⅳ. ① TS972.123

　　中国版本图书馆CIP数据核字 (2015) 第 005995 号

我爱中国菜·素食篇
心蓝　编著

©华语教学出版社有限责任公司
华语教学出版社有限责任公司出版
（中国北京百万庄大街 24 号 邮政编码 100037）
电话：(86)10-68320585 68997826
传真：(86)10-68997826 68326333
网址：www.sinolingua.com.cn
电子信箱：hyjx@sinolingua.com.cn
唐山玺诚印务有限公司印刷
2015 年（32 开）第 1 版
2024 年第 1 版第 3 次印刷
（英文版）
ISBN 978-7-5138-0929-0
003900